INCREDIBLE ANIMAL FACE-OFFS

★ ANIMAL ★
SPEED
SHOWDOWN

ELSIE OLSON

Consulting Editor, Diane Craig, M.A./Reading Specialist

Super Sandcastle

An Imprint of Abdo Publishing
abdobooks.com

abdobooks.com

Published by Abdo Publishing, a division of ABDO, PO Box 398166, Minneapolis, Minnesota 55439. Copyright © 2020 by Abdo Consulting Group, Inc. International copyrights reserved in all countries. No part of this book may be reproduced in any form without written permission from the publisher. Super Sandcastle™ is a trademark and logo of Abdo Publishing.

Printed in the United States of America, North Mankato, Minnesota
102019
012020

Design: Sarah DeYoung, Mighty Media, Inc.
Production: Mighty Media, Inc.
Editor: Jessica Rusick
Cover Photographs: Shutterstock Images
Interior Photographs: Getty Images/iStockphoto, pp. 11, 12, 13; Shutterstock Images, pp. 4, 5, 6, 7, 8, 9, 10, 13 (right), 14, 15, 16, 17, 18, 19, 20, 21, 22, 23

Library of Congress Control Number: 2019943210

Publisher's Cataloging-in-Publication Data
Names: Olson, Elsie, author.
Title: Animal speed showdown / by Elsie Olson
Description: Minneapolis, Minnesota : Abdo Publishing, 2020 | Series: Incredible animal face-offs
Identifiers: ISBN 9781532191978 (lib. bdg.) | ISBN 9781532178771 (ebook)
Subjects: LCSH: Movements of animals (Physiology)--Juvenile literature. | Animals, Habits and behavior of-
    -Juvenile literature. | Social behavior in animals--Juvenile literature. | Animal defense mechanisms—
    Juvenile literature.
Classification: DDC 591.57--dc23

Super Sandcastle™ books are created by a team of professional educators, reading specialists, and content developers around five essential components—phonemic awareness, phonics, vocabulary, text comprehension, and fluency—to assist young readers as they develop reading skills and strategies and increase their general knowledge. All books are written, reviewed, and leveled for guided reading, early reading intervention, and Accelerated Reader™ programs for use in shared, guided, and independent reading and writing activities to support a balanced approach to literacy instruction.

CONTENTS

BATTLE OF THE SPEEDERS

The animal kingdom is full of stars. But some animals stand out. These animals are the fastest of their kinds.

Some animals use speed to travel long distances. Others use speed to escape predators.

Speedy creatures are all around us. But what if you matched them up in face-offs? Which animal would win a speed showdown?

PEREGRINE FALCON

PRONGHORN

SQUID

SAILFISH

OSTRICH

CHEETAH

5

HUNTERS GO HEAD-TO-HEAD

Small, sleek body

Cheetahs are high-speed hunters. Peregrine falcons are too! But which animal would win if these hunters went head-to-head?

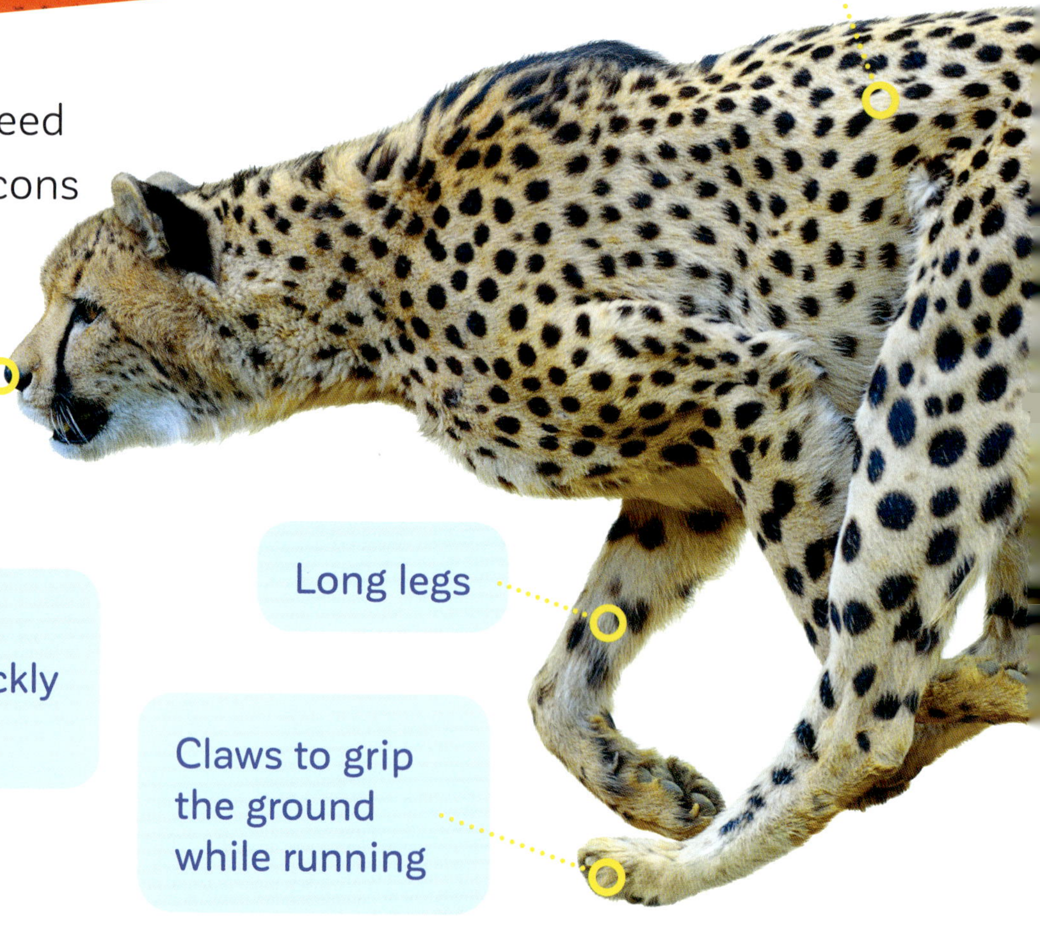

Long legs

Large **nostrils** to breathe in air quickly while running

Claws to grip the ground while running

Tail to help steer while running

Spotted fur to hide in tall grass

CHEETAH
BORN TO RUN

This big cat could outrun your school bus while chasing down a meal. Put your hands together for the astonishing cheetah!

CHEETAH STATS

HOME
Grasslands in Africa and parts of Asia

FOOD
Antelope, ostriches, small birds, and small **mammals**

SIZE
Up to 7 feet (2 m) long and 145 pounds (66 kg)

A CHEETAH IS LARGER THAN A PET LABRADOR.

CHEETAH

LABRADOR

Strong chest muscles for flapping wings

Sleek build

PEREGRINE FALCON
FAST FLYER

This swift diver is the fastest animal on Earth. Raise your voice for the powerful peregrine falcon!

Long wingspan

Sharp **talons** for striking prey

PEREGRINE FALCON STATS

HOME
In cities, grasslands, and along shorelines of every continent but Antarctica

FOOD
Other birds, including pigeons, ducks, and gulls

SIZE
Up to 19 inches (48 cm) with a 4-foot (1 m) wingspan

A PEREGRINE FALCON IS LARGER THAN AN AMERICAN ROBIN.

PEREGRINE FALCON

AMERICAN ROBIN

CHEETAH VS PEREGRINE FALCON

CHEETAH

Cheetahs use speed to hunt. But they can only run at top speed for so long. So, they must be smart!

PREDATOR POWER

Cheetahs hide in tall grass. They sneak up on prey. Once prey is within 165 feet (50 m), the cheetah charges!

THAT'S SPEEDY!

A cheetah can run up to 75 miles per hour (121 kmh). Cheetahs can hold this speed for 30 seconds.

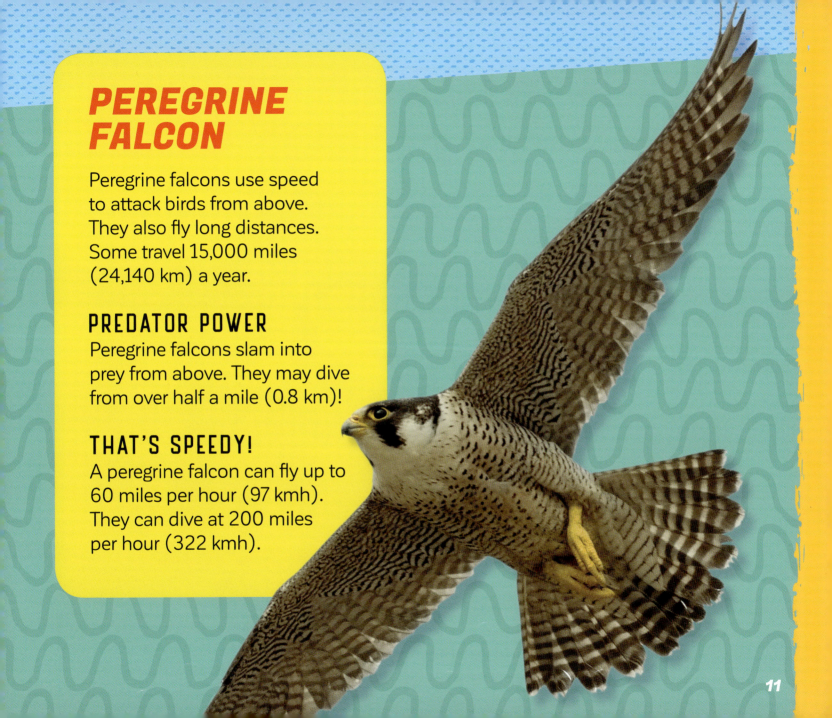

PEREGRINE FALCON

Peregrine falcons use speed to attack birds from above. They also fly long distances. Some travel 15,000 miles (24,140 km) a year.

PREDATOR POWER

Peregrine falcons slam into prey from above. They may dive from over half a mile (0.8 km)!

THAT'S SPEEDY!

A peregrine falcon can fly up to 60 miles per hour (97 kmh). They can dive at 200 miles per hour (322 kmh).

LONG-DISTANCE DUEL

Pronghorns are excellent endurance runners. Ostriches are too! But which animal would win a long-distance duel?

Horns to defend against predators

Large eyes can spot predators up to three miles (5 km) away

Long, slender legs

PRONGHORN
PREDATOR ESCAPE ARTIST

This speedy **sprinter** leaves its predators in the dust. Get on your feet for the swift pronghorn!

White patch on backside to signal herd members

PRONGHORN STATS

HOME
Grasslands and deserts in North America

FOOD
Plants, including weeds and grasses

SIZE
Up to 5 feet (1.5 m) long and 150 pounds (68 kg)

A PRONGHORN IS SMALLER THAN MOST ADULT BIKES.

PRONGHORN

BIKE

Long, strong legs help an ostrich stride up to 16 feet (5 m)

Each foot has two toes with sharp claws for gripping the ground

Wings to help change direction while running

OSTRICH
FLIGHTLESS BUT FLEET-FOOTED

This big bird is flightless. But it isn't slow! Make some noise for the outstanding ostrich!

OSTRICH STATS

HOME
African deserts and grasslands

FOOD
Plants, seeds, insects, and small reptiles

SIZE
Up to 9 feet (2.7 m) tall and 350 pounds (159 kg)

AN OSTRICH IS TALLER THAN AN AVERAGE MAN.

OSTRICH   MAN

PRONGHORN VS OSTRICH

PRONGHORN

Pronghorns live in open spaces. There is nowhere to hide! So, they use speed to outrun predators.

GREAT ESCAPERS

A pronghorn raises its tail when it sees a predator. This reveals a white patch. The patch is a warning. It tells the pronghorn herd to run!

THAT'S SPEEDY!

Pronghorns can **sprint** at 60 miles per hour (97 kmh) for short distances. They can run long distances at 30 miles per hour (48 kmh).

OSTRICH

An ostrich's kick can kill a lion! But this bird usually uses its legs to run.

GREAT ESCAPERS
Predators can **sprint** faster than an ostrich. But ostriches have greater endurance. With a head start, ostriches can escape anything!

THAT'S SPEEDY!
Ostriches can sprint at 40 miles per hour (64 kmh) for short distances. They can run long distances at 30 miles per hour (48 kmh).

UNDERWATER RACE

Hollow **mantle** to store water

Squid are speedy swimmers. Sailfish are too! But which animal would win an underwater race?

SQUID
DEEP-WATER SPEEDSTER

This cool creature uses **propulsion** to speed through the ocean. Meet the superfast squid!

Siphon to push water out and control speed and direction

Large eyes to see light in deep water

Thin, sleek body

Eight arms for swimming

Two **tentacles** for snatching prey

SQUID STATS

HOME
Deep water in oceans all over the world

FOOD
Fish, crabs, shrimp, and other sea creatures

SIZE
Less than 1 inch (2.5 cm) long to 43 feet (13 m) long, depending on species

A GIANT SQUID IS ALMOST AS LONG AS A SCHOOL BUS.

GIANT SQUID

SCHOOL BUS

Sail-like fin to
block and trap prey

Large gills for
taking in oxygen

Long, spear-like bill
for striking prey

SAILFISH
PHENOMENAL FISH

This fast fish is one of the ocean's top hunters. Say hello to the marvelous sailfish!

Long, sleek body

SAILFISH STATS

HOME
Open oceans all over the world

FOOD
Fish, squid, octopus, and other ocean animals

SIZE
Up to 10 feet (3 m) long and 220 pounds (100 kg)

A SAILFISH IS SHORTER THAN A CAR.

SAILFISH

CAR

SQUID VS SAILFISH

SQUID

Squid are unlike most other ocean animals. This is because they swim tail-first!

FAST AND FURIOUS

Squid use speed to escape whales and sharks. Squid can't stop swimming. If they do, they sink!

THAT'S SPEEDY!

Squid swim by taking in water. Then, they push the water out. This force moves the squid. They can reach speeds up to 25 miles per hour (40 kmh)!

SAILFISH

Sailfish hunt in groups. They quickly swing their bills side to side. This stuns or kills prey!

FAST AND FURIOUS

Sailfish swim in short, quick bursts. While hunting, sailfish leap into the air. This leaping helps force prey into one area.

THAT'S SPEEDY!

Sailfish can jump at 68 miles per hour (109 kmh). Scientists believe they are the ocean's fastest fish!

GLOSSARY

mammal—a warm-blooded animal that has hair and whose females produce milk to feed their young.

mantle—an external part of the body that covers a squid's internal organs.

nostril—an opening in the nose.

propulsion—something that drives forward or adds speed to an object.

siphon—a tube-shaped organ for drawing in and expelling liquids.

sprint—to run as fast as possible for a short distance.

talon—the claw of an animal, especially that of a bird of prey.

tentacle—a long, flexible limb on an invertebrate such as a jellyfish or squid.